DATA
SIMPLIFIED

A Guide for All on Making Sense of

the Complex Terms in Technology

Arin Tahmasian

Who Should Read This Book?

This book is a perfect match for anyone eager to peel back the layers on how data shapes our technology-driven world. Whether you're just curious about how data is structured, stored, shared, and utilized, or you're diving into the specifics of data's role in technology, this guide has something for you. While having some familiarity with software concepts might smooth your journey through the pages, it's not a prerequisite. We start with the basics, ensuring that no prior knowledge of data structures or data storage is needed to get started.

As we explore the fundamentals of data together, you'll gain a clear and thorough understanding of its role in our digital age.

My Journey and Mission

I am Arin Tahmasian, an engineer with over three decades in the software industry. I have dedicated over twenty years to assisting small businesses in meeting their software needs. The evolution of technology, from personal computers and internet networks to cloud computing solutions and AI, is a journey that I have actively participated in. I believe everyone has a unique contribution to make to humanity and our world. Sharing my insights and fostering understanding in technology is my way of making a difference.

Setting Sail on a Journey Together

From as early as I can remember, science and technology have not just been interests but passions that define who I am. My belief has always been that true learning comes from a deep, fundamental understanding of a subject. This philosophy, combined with my love for teaching, inspired me to create the "Tech Simplified" book series. My goal is to demystify complex technological concepts, presenting them in a manner that is both simple and fundamental.

This approach to learning has been incredibly rewarding for me, and I'm excited to share it with you. I invite you to join me on this journey of discovery. Whether you're a beginner or someone looking to deepen your understanding, "Tech Simplified ™" is designed to make learning about technology both fun and practical.

As you turn these pages, I hope you find the same joy and curiosity in reading this book as I did in creating it. And for those eager to explore further, don't forget to check out my other books in the series – simply scan the QR code provided below for more enlightening adventures into the fundamentals of technology.

Welcome aboard!

Arin Tahmasian

Author, and Founder of "Tech Simplified ™" Book Series

Contents

If you can't explain it simply, you don't understand it well enough.

– Albert Einstein

PART I

The ABCs

of

Data

The Foundation

A Brief History Behind the Word 'Data'

Once upon a time, in the world of words and their winding journeys, "data" began its life as a humble Latin term. The story of "data" is a tale of transformation, from ancient inscriptions to the heart of today's digital age.

Imagine a Roman scholar, parchment in hand, using the word "datum" (meaning "something given") to refer to a fact laid before him, a piece of evidence to ponder or debate. In its earliest days, "datum" was the singular form, a lone fact standing in the light of scholarly examination. As centuries passed, this Latin word traveled through time, crossing languages and borders, evolving in form and pluralizing to "data," a collection of facts waiting to be discovered, analyzed, and interpreted.

Fast forward to the Age of Enlightenment, where the seeds of modern science and reasoning were sown. Here, "data" found fertile ground, becoming the cornerstone of empirical evidence and scientific discovery. Philosophers and scientists gathered data like precious gems, using it to build the edifice of modern knowledge.

As the industrial era dawned, the word "data" began to dress in new clothes, stepping into the bustling streets of the 19th and 20th centuries. The invention of the telegraph, the telephone, and eventually the computer transformed "data" from static facts on paper to dynamic bits of information, pulsing through wires and lighting up screens.

In today's digital world, "data" has swelled into an ocean that envelops our lives. With every click, swipe, and key-stroke, we contribute to the ever-expanding universe of digital data. From social media posts to satellite images, data has become the lifeblood of decision-making, the fuel for innovation, and the canvas for creativity in the 21st century.

The journey of "data" is a mirror reflecting our quest for knowledge, our desire to understand the world around us, and our efforts to shape the future. From a single "datum" in the hands of ancient scholars to the vast digital landscapes of today, "data" has woven itself into the fabric of human history, reminding us that our quest for understanding and the stories we tell are as infinite as the data points that dot the cosmos.

The Concept of Data in Our Modern World

In the tapestry of our modern society, data threads through every aspect, weaving patterns that shape our understanding, decisions, and innovations. Far from its humble origins as discrete bits of information, data has grown into a force that's everywhere, driving the pulse of our digital ecosystem.

At the individual level, data plays a pivotal role in shaping experiences and behaviors. From personalized recommendations on streaming platforms to tailored advertisements on social media, data analysis enables a customization of life's digital interface, making our interaction with technology both seamless and intuitive. Moreover, wearable technology and health apps collect data to provide insights into our well-being, offering a personalized blueprint for health and fitness.

In the world of business and industry, data is the foundation of strategic decision-making. Big data analytics, a field that has gained momentum over the last decade, allows companies to harness vast amounts of information to optimize operations, understand consumer behavior, and predict market trends. This data-driven approach has revolutionized industries, from retail and finance to healthcare and transportation, enabling efficiency and innovation previously unimaginable.

The influence of data extends into the public sector, where it serves as a critical tool for governance and public service delivery. Governments around the world utilize data to enhance urban planning, manage public health, and ensure national security. The use of data in managing the COVID-19 pandemic, through tracking infection rates and vaccine distribution, underscores its vital role in crisis management and public health strategies.

Yet, the ubiquity of data brings with it challenges and ethical considerations. Privacy concerns, data security, and the digital divide are pressing issues that societies must address. The question of who owns data, how it is used, and the implications for privacy and autonomy are central to the discourse on data in the modern world.

Furthermore, the concept of "big data" has introduced a paradigm shift in how we perceive information. The ability to process and analyze data sets of unprecedented size and complexity has opened new frontiers in research, innovation, and technology development. Fields such as artificial intelligence and machine learning are predicated on the availability of large-scale data, driving advancements that have the potential to reshape every aspect of human life.

Bits – The Foundation of Digital Data

At the core of our computer systems lies the binary concept. This fundamental principle dictates that every piece of information is represented as a series of 0s and 1s. When computers were first developed, the ingenious idea of combining these binary digits, or "bits," into meaningful sequences laid the groundwork for digital data.

Consider the "bit" as the most basic unit of data in the computer world. You've likely encountered the term "Megabit Per Second" when evaluating internet services, indicating the millions of bits that can be transmitted to your computer every second.

A single bit, capable of holding a value of either 0 or 1, can convey one of two possible states: true or false. This binary approach allows for straightforward, yet profoundly effective, data representation. For instance, to store a simple yes or no answer, a single bit suffices. However, the real magic unfolds when we start to combine bits.

Imagine we have a pair of bits. The combinations they can form are as follows:

```
0, 0
0, 1
1, 0
1, 1
```

With just two bits, we can represent four distinct outcomes, demonstrating that combined bits can hold more information together than individually. This principle is the cornerstone of digital data in modern technology.

Expanding our view to three bits, let's map each possible combination to a familiar decimal number, starting from zero:

```
0: 0,0,0
1: 0,0,1
2: 0,1,0
3: 0,1,1
4: 1,0,0
5: 1,0,1
6: 1,1,0
7: 1,1,1
```

By adding just one more bit, we've doubled the potential data storage capacity. This illustrates the exponential power of the binary system:

With 2 bits, we can store 4 distinct data outcomes (2^2).

With 3 bits, the capacity increases to 8 outcomes (2^3).

This doubling effect with each additional bit highlights the efficiency and scalability of binary data storage in computer systems. Through this simple yet powerful system, we unlock the vast potential of digital technology, laying the foundation for the complex computations and data storage mechanisms that drive our world today.

ASCII (American Standard Code for Information Interchange)

In the 1960s, the development of ASCII, pronounced 'askie,' marked a significant milestone in digital communication. ASCII introduced a collection of 128 characters, encompassing both letters and numbers. To accommodate one ASCII character, 7 bits were required. The adoption of an 8-bit byte facilitated the representation of all ASCII characters and also provided an additional bit that could be used for extending the character set or implementing error checking mechanisms.

This advancement led to the standardization of the 8-bit *byte*, which became a foundational element in the architecture of modern computer systems. The introduction of the byte revolutionized data storage, enabling computers to process and store information in a format that is easily readable and meaningful to humans.

Here's a glimpse into how ASCII codes work: each letter of the English alphabet is mapped to a specific byte that represents the letter. But ASCII isn't just about letters; it also encompasses symbols, numbers, and special characters. For instance, the 'CR' or Carriage Return, which corresponds to the 'Enter' key on your keyboard, is part of the ASCII set. This system ensures that common characters and commands have a universal binary representation, facilitating standardized digital communication.

Decimal	Binary	Octal	Hex	ASCII	Decimal	Binary	Octal	Hex	ASCII
32	00100000	040	20	SP	64	01000000	100	40	@
33	00100001	041	21	!	65	01000001	101	41	A
34	00100010	042	22	"	66	01000010	102	42	B
35	00100011	043	23	#	67	01000011	103	43	C
36	00100100	044	24	$	68	01000100	104	44	D
37	00100101	045	25	%	69	01000101	105	45	E
38	00100110	046	26	&	70	01000110	106	46	F
39	00100111	047	27	'	71	01000111	107	47	G
40	00101000	050	28	(	72	01001000	110	48	H
41	00101001	051	29	)	73	01001001	111	49	I
42	00101010	052	2A	*	74	01001010	112	4A	J
43	00101011	053	2B	+	75	01001011	113	4B	K
44	00101100	054	2C	,	76	01001100	114	4C	L
45	00101101	055	2D	-	77	01001101	115	4D	M
46	00101110	056	2E	.	78	01001110	116	4E	N
47	00101111	057	2F	/	79	01001111	117	4F	O
48	00110000	060	30	0	80	01010000	120	50	P
49	00110001	061	31	1	81	01010001	121	51	Q
50	00110010	062	32	2	82	01010010	122	52	R
51	00110011	063	33	3	83	01010011	123	53	S
52	00110100	064	34	4	84	01010100	124	54	T
53	00110101	065	35	5	85	01010101	125	55	U
54	00110110	066	36	6	86	01010110	126	56	V
55	00110111	067	37	7	87	01010111	127	57	W
56	00111000	070	38	8	88	01011000	130	58	X
57	00111001	071	39	9	89	01011001	131	59	Y
58	00111010	072	3A	:	90	01011010	132	5A	Z
59	00111011	073	3B	;	91	01011011	133	5B	[
60	00111100	074	3C	<	92	01011100	134	5C	\
61	00111101	075	3D	=	93	01011101	135	5D	]
62	00111110	076	3E	>	94	01011110	136	5E	^
63	00111111	077	3F	?	95	01011111	137	5F	_

Figure 1 – ASCII Code Sample

Now, by assembling a series of bytes, we can store meaningful information within a computer system.

For example, consider how the word 'DATA' is represented using bytes:

```
D: 01000100
A: 01000001
T: 01010100
A: 01000001
```

With this foundational understanding, let's leap forward to the capabilities of contemporary computers, which can process and store millions of bytes in mere seconds.

Data Repository

In the world of computing, a data repository refers to any system capable of holding vast quantities of data. One of the most familiar forms of such repositories is the computer's file system. This system is where all your digital belongings, like photos, music, applications, and more, are stored. The introduction of storage units necessitated a standardized method for data storage within computer systems, leading to the development of file systems. These systems provide a uniform way to organize data on your computer's hard drives (or other storage devices). A file represents a coherent collection of data that computer systems can read, interpret, and present to the user in a meaningful way.

Data Types and Structures

In digital computing, data types and structures are crucial for defining the nature of data and how it can be manipulated within a program or a computing environment. Here's an overview of fundamental data types and structures that are foundational in programming and data processing.

Fundamental Data Types

Integer

Represents whole numbers, both positive and negative. It does not include decimal points and can be used for counting or indexing arrays.

Example: 24,68,9867

Floating Point

Used to represent real numbers, including fractions or numbers with decimal points. It's crucial for any calculations requiring precision beyond whole numbers.

Example: 2.365, 3.1415, 8654.876

Character

Represents single alphanumeric characters or symbols, typically stored in ASCII or Unicode format. It's used for text processing.

Example: 'L' , 'a' , 'T'

String

A sequence of characters treated as a single unit. Strings are used to store text, such as names or sentences.

Example: "This is a string!"

Boolean

A simple data type with only two possible values: true or false. Booleans are essential for conditional statements and logic in programming.

Example: "True" , "False"

Composite Data Types

A Composite Data Type, also known as a compound or complex data type, refers to a data type that is composed of two or more primitive (simple) data types or other composite types. Unlike primitive data types such as integers, booleans, and characters, which can hold only one value at a time, composite data types can aggregate multiple values, possibly of different types, and manage them as a single unit. This capability makes them indispensable for representing more complex data structures in programming and computing.

Examples of Composite Data Types

Arrays

An array is a composite data type that consists of a collection of elements, all of the same type, stored in contiguous memory locations. The elements can be accessed by their index in the array.

Example: [4, 7, 87, 32, 25]

Structures (Structs)

Structures are composite types that allow for the grouping of variables of different types under a single name. For example, a struct could represent a Person with attributes like name (string), age (integer), and height (float).

Example: John = { "name": "John", "age": 27, "height": 68}

Classes

Classes are more complex composite types that encapsulate data (attributes) and methods (functions) that operate on the data. They are a fundamental concept in object-oriented programming, allowing for the creation of objects.

Example:

```
Class Book
    Initialize(title, author, pages)
        Set self.title to title
        Set self.author to author
        Set self.pages to pages

    Method display_info
        Print "Book: " + self.title
        Print "Author: " + self.author
        Print "Pages: " + self.pages

End Class
```

Lists

Lists (or linked lists) are collections of items where each item can be of a different type. They are dynamic, meaning they can grow or shrink in size, and each element points to the next, allowing for efficient insertion and deletion.

Example:

```
// Create a List with different data types
List mixedList

// Add items of various types to the List
Add 42 to mixedList            // Integer
Add "Hello, World!" to mixedList  // String
Add True to mixedList          // Boolean
Add 3.14 to mixedList          // Floating-point num-
ber
```

Tuples: Tuples are similar to arrays but can contain elements of different types. They are immutable, meaning their values cannot be changed once set.

Example:

my_tuple = (32, "Apple", True, 3.14)

Specialized Data Structures

A data structure is a particular way of organizing, managing, and storing data in a computer so that it can be accessed and modified efficiently. Essentially, data structures are the building blocks for data storage and manipulation, providing a means to manage large amounts of data effectively for various purposes, such as indexing, searching, and sorting. They are crucial for creating efficient algorithms and software applications.

Linked Lists

A collection of items where each item points to the next one, allowing for dynamic insertion and removal of elements.

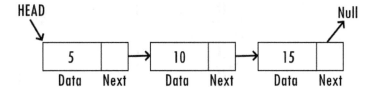

In this example, 5 points to 10, and 10 points to 15, and 15 points to nothing which in computer world is called "null".

A linked list can be either singly or doubly linked. A node in a singly linked list contains a reference or pointer to the next item, facilitating forward traversal through the list. Conversely, a node in a doubly linked list includes references to both the next and the previous items. This dual referencing allows traversal in both forward and backward directions, enhancing navigation flexibility. In essence, while each node in a singly linked list is aware of its successor, nodes in a doubly linked list are aware of both their preceding and following counterparts.

Stacks

Operates on a Last In, First Out (LIFO) principle. Useful for situations where you need to reverse actions or navigate back to an initial state.

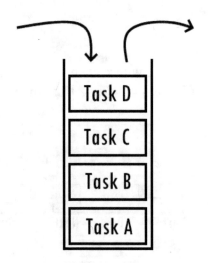

Visualize a stack as a bucket where you can place your to-do list items. Unlike a queue, in this bucket, the first task you add will be the last one you take out.

This concept is particularly valuable in the computing world for handling nested operations. Here, tasks are stacked from the outermost to the innermost. Consequently, the computer addresses these tasks in reverse order, tackling the most recently added task first and progressing to the earliest task. This 'Last In, First Out' (LIFO) approach is crucial for efficiently managing a sequence of operations.

Queues

Operates on a First In, First Out (FIFO) principle, ideal for tasks that need to be processed in the order they were added.

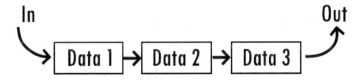

Trees

Imagine a tree in nature: it has a trunk, branches, and leaves. In the world of computers, we have something quite similar called a "tree" data structure, but instead of leaves and branches, it holds pieces of information.

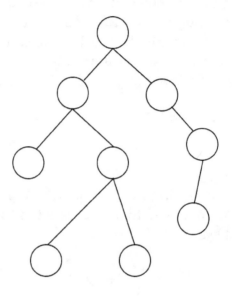

Here's a simple way to understand it...

The Trunk

Think of the trunk of the tree as the starting point. In computer terms, we call this the "root." It's the very top item of the tree from where everything begins.

Branches

Just as branches grow out from the trunk, in a tree data structure, we have branches that stem from the root. These branches can split off into more branches, allowing us to organize information at various levels.

Leaves

At the ends of the branches, instead of leaves, we have nodes (you can think of them as the final pieces of information). These nodes don't have any branches coming out of them; they're the endpoints.

One of the most special things about a tree in computing is that it can be used to store information that naturally forms a hierarchy. For example, a family tree is a great real-world example of a tree structure, with each generation branching off the previous one.

Trees are super useful for organizing things in a way that makes it easy to find what you're looking for, much like how branches and leaves are arranged in a natural tree. For instance, they're used in computer programs to manage files on your computer, help play chess by looking ahead at different moves, or even organize web pages in your internet browser history.

Graphs

Imagine a bustling city map, with places connected by a network of roads. In computing, we have something quite similar known as a "graph" data structure, which acts like a web of points linked together by lines.

In this city of data, the points are like various landmarks or destinations, which we call "vertices" in graph terms. Each vertex is a piece of data or an item in our digital landscape. The roads that connect these places, allowing you to travel from one spot to another, are the "edges" in our graph. These edges represent the relationships or paths between our data points, showing how they're interconnected.

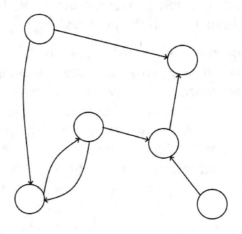

Think about how diverse the connections in a city can be. Some roads might be one-way, representing a direction in the relationship, while others are two-way streets, indicating a bidirectional connection. This flexibility in mapping out connections is what makes graphs incredibly useful for representing complex networks in computing.

Graphs come to life in various real-world scenarios, far beyond city maps. In social networks, each person is a vertex, and friendships form the edges, creating a vast network of connections. On the internet, websites become the vertices, with hyperlinks as edges weaving a complex web of online interactions. Even in transportation, airports and the flights between them can be modeled as graphs, helping to navigate the global network of air travel.

Unlike trees, which have a clear hierarchy from root to leaves, graphs thrive on their ability to illustrate a maze of interconnections without strict hierarchies. They can include cycles, like roundabouts, letting you explore and loop back, and they offer multiple pathways from one point to another, mirroring the complexity and richness of real-world interactions.

So, when we talk about a graph data structure, we're looking at a powerful way to depict and navigate the intricate relationships between data, capturing the essence of how items or entities are connected in a network that's as dynamic and complex as the streets of a city.

Hash Tables

Imagine walking into a vast library filled with thousands of books, and you have the magical ability to find any book instantly, regardless of how many are there. This is what a hash table does in the world of computing.

At its core, a hash table acts like the library's advanced cataloging system. Instead of combing through every book one by one, you use a unique identifier for each book, like its ISBN, to pinpoint its exact location on the shelves. This process involves a special formula known as a *hash function*, which takes the book's identifier and calculates a unique spot, or "index," where this book belongs in the library.

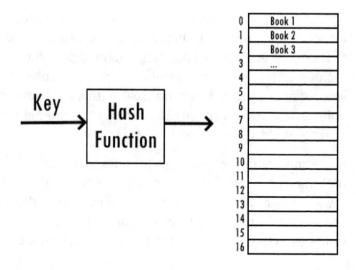

Now, whenever you need to find a book, you apply the same formula, and like magic, it directs you to the exact location of the book. This method allows you to bypass the tedious process of searching aisle by aisle and go directly to where the book is stored. It's like having a magic map in your hands.

However, there's a twist. Sometimes, two books might be assigned the same spot, a dilemma known as a *"collision."* But the hash table is equipped to handle such situations gracefully, perhaps by designating a secondary location for the second book or linking it to the first, ensuring that both can be found without hassle.

The brilliance of hash tables lies in their efficiency. Regardless of the size of the data, hash tables maintain their ability to access, add, or remove items with remarkable speed. This efficiency makes them an indispensable tool for programmers who deal with large volumes of data and require a method to manage it swiftly and effectively.

Before we wrap up this chapter, there's one more essential concept we need to explore: Algorithms in data processing. This key topic will shed light on how data is transformed and utilized effectively.

Understanding Algorithms

Imagine you have a recipe for baking a cake. This recipe provides step-by-step instructions on what ingredients you need, how to mix them, and how long to bake them to end up with a delicious cake. In computing, an algorithm works similarly—it's a recipe that tells the computer how to achieve a particular outcome from a given set of inputs.

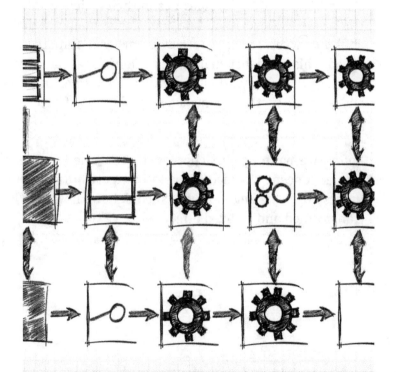

Algorithms play a crucial role in the digital world for a multitude of reasons. Firstly, they are the backbone of efficiency in computing, determining the fastest and most resource-effective methods for processing data, solving problems, and carrying out tasks. Moreover, they are instrumental in problem-solving, whether it's for conducting searches on the internet or performing intricate scientific calculations. Behind every query and computation, algorithms work tirelessly, making these processes feasible. Additionally, they are key to decision-making within software applications, automating choices ranging from movie recommendations on streaming platforms to spotting trends within datasets.

Types of Algorithms

In the domain of data processing, a myriad of algorithms exist that significantly enhance the speed and efficiency with which we access and manage data. These algorithms have undergone considerable evolution, greatly improving our ability to search through vast datasets and optimizing power consumption for data storage and retrieval. In this section, we will explore some of the most renowned search and sorting algorithms, offering insights into how these algorithms operate to deliver the requested data efficiently.

Sorting Algorithms

Before diving into famous sorting algorithms, let's grasp the significance of sorting data. To put it simply, whether it's our work desk in real life or our closet, an organized environment naturally enhances efficiency and speed. Consider the closet analogy. Imagine having a cluttered closet where shirts, pants, socks, and t-shirts are all jumbled together. Finding a specific item in such chaos would be challenging. Now, envision a closet where everything is neatly organized by category. Life suddenly becomes much simpler. Taking it a step further, if items are sorted not only by type but also by color, locating that blue t-shirt you were looking for becomes a breeze. The principle is the same with computer systems and data. When data is unsorted in a repository, the computer must scan the entire dataset to locate the desired information. Searching through millions of bytes of data could be painstakingly slow. However, with sorted data, the system can efficiently pinpoint exactly where to look, significantly speeding up data retrieval.

Drawing from personal experience, I once tackled a challenge to enhance database data retrieval speeds. The customer had complained about slow data retrievals that bogged down the system whenever they searched for specific items. By sorting the data and properly indexing it, what previously took 10 minutes was reduced to merely 4 seconds. This optimization not only expedited the process but also introduced computational efficiency, saving both time and resources.

Now, let's explore a few sorting algorithms and dive into their mechanisms...

Bubble Sort

Imagine you have a line of numbered balloons from 1 to 10, but they're all mixed up, and you want to arrange them in order from smallest to largest. Bubble Sort is like a gentle breeze that repeatedly sweeps through the balloons, comparing each pair of neighboring balloons and swapping them if they're in the wrong order. After each pass, the highest numbered balloon in the unsorted section floats to its correct position at the end, just like the biggest bubble rising to the surface of water.

Let's look at an example of sorting the following list: [8, 7, 5, 4] in ascending order.

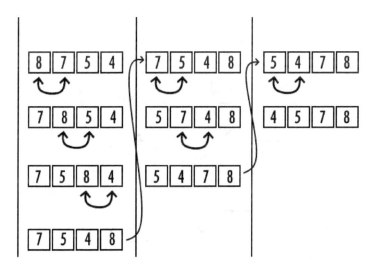

Merge Sorting

Imagine you're organizing a deck of shuffled playing cards into order. Merge Sort is like a clever strategy of sorting that deck. Instead of trying to sort the entire deck at once, you divide it into smaller and smaller groups until each group has just one card. Then, you start combining these single-card groups back together, but as you do, you make sure each combined group is sorted. It's like sorting two tiny, easy-to-manage piles of cards and then putting them together into one sorted pile, and so on, until the whole deck is back together and sorted.

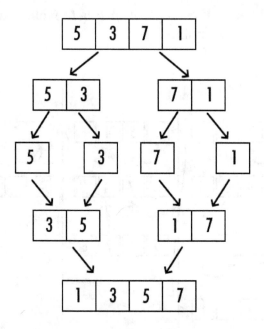

Merge Sort is efficient and particularly good for large lists because it systematically organizes smaller sections, making the task of sorting the whole much more manageable.

Quick Sort

Imagine you're the captain of a team picking players for a dodgeball game from a line-up. Quick Sort is like a smart strategy to organize players by height quickly, so it's easier to see who's tallest and who's not. Here's how you might do it:

Step 1: Pick a Pivot

First, you choose one player to compare others against. Let's say you pick someone from the middle of the line. This player is called the "pivot."

Step 2: Divide the Line

Next, you ask all players shorter than the pivot to move to the left side and everyone taller to move to the right side. Now, without sorting everyone precisely, you've quickly created two groups based on height.

Step 3: Sort the Groups

Now, imagine each group forms its own smaller line. You repeat the process for these lines: pick a new pivot in each, divide again into taller and shorter, and so on. Each time, the groups get smaller and more manageable.

Step 4: Repeat Until Sorted

You keep doing this - picking pivots, dividing groups - until each small group has just one player or none. By now, everyone is in place as if they've been sorted by height from shortest to tallest.

Quick Sort is like this game of dividing players into groups and sorting those groups. It's quick because, at each step, you reduce the problem size by creating smaller groups and sorting those. It's a clever way to sort without having to compare every player to every other player directly, making it a fast method to get everyone organized.

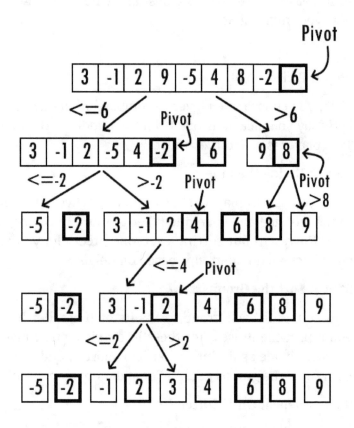

Search Algorithms

Now, let's turn our attention to search algorithms. Searching is an essential activity we engage in regularly, from seeking out a book on a library shelf to finding a contact's name on our phones. In the digital world, search algorithms play a pivotal role, making it possible to swiftly and effectively locate specific pieces of data within extensive datasets.

Linear Search

Imagine you've lost your favorite sock in a drawer full of clothes. To find it, you start at one end of the drawer and rummage through each item, one by one, until you find your sock. This process is very much like the Linear Search algorithm in computing.

Linear Search is straightforward and doesn't require the list to be in any particular order, making it versatile but not the fastest method, especially for large lists. It's best used for small lists or when the list's order is unknown.

Binary Search

Imagine you're playing a guessing game where you have to find a number between 1 and 100, and after each guess, you're told whether the actual number is higher or lower. Instead of guessing numbers randomly, you decide to start with 50. If the number is higher, you then guess 75, cutting down the range of possibilities significantly with each guess. This strategy is similar to how Binary Search works.

Binary Search is efficient and fast, especially for large lists, but it requires the list to be sorted beforehand. It's like the guessing game where strategy and elimination quickly lead you to the correct number—or in the case of Binary Search, the item you're looking for.

With that, we conclude this chapter. You now possess a solid foundation in the basics of data within the digital realm. In the forthcoming chapter, we will explore the security aspects of data, examining strategies for safe-guarding it both in transit across networks and while at rest on storage devices such as hard drives.

PART II

Data Security & Redundancy

The Fortress

Having gained an understanding of data and its storage and management in the digital world, we now shift our attention to the critical aspect of data security. Just as we protect our valuables in the physical realm, data in the digital domain demands even greater vigilance to safeguard our digital assets. Often, a security breach may not be as immediately noticeable or preventable as incidents in the physical world. Consequently, significant effort is directed towards preventing and safeguarding against digital data attacks, rather than merely responding to them after they occur.

This chapter will concentrate on data security, examining the measures in place to safeguard data both during its transfer between systems and when stationary on hard drives within data centers. Additionally, we will briefly explore the mechanisms that ensure data integrity within storage units. Finally, we will discuss various techniques employed to achieve data redundancy.

The Importance of Data Security

Thanks to advances in communication technology, along with the evolution of cloud computing and hardware, access to data in the digital world has grown exponentially. This surge in computational power, coupled with the increasing volume of data stored in the digital cloud, highlights the critical importance of focusing on a somewhat silent yet immensely significant topic: data security. Now, more than ever, it's crucial to understand why securing data in our digital age is paramount. Let's explore the reasons that make data security indispensable in today's digital landscape.

Privacy Protection

Our personal information, ranging from social security numbers to bank account details, is continuously at risk of falling into the wrong hands. Data security measures are essential to protect this information, thereby preserving individual privacy and preventing identity theft. Without robust data security, personal details could be exploited, leading to fraud, financial loss, and a significant invasion of privacy.

Maintaining Trust

For businesses, the security of customer data is a cornerstone of customer trust and brand integrity. In an era where data breaches are increasingly common, a single incident can have far-reaching consequences. A breach can lead to substantial financial losses, erode customer confidence, and inflict lasting damage on a company's reputation. Securing data effectively is therefore crucial for maintaining the trust that customers place in a business.

Regulatory Compliance

Data security also intersects significantly with legal and regulatory requirements. Various industries are governed by regulations that dictate how sensitive data must be handled and protected. These regulations, such as GDPR in the European Union or HIPAA in the United States, aim to ensure that organizations take the necessary precautions to protect consumer and patient information. Non-compliance with these regulations can lead to severe penalties, including hefty fines and legal action, underscoring the importance of adhering to data security standards.

Data Encryption

Encryption stands as the guardian of privacy in our digital communications, transforming readable data into a coded form that can be decrypted only by those who possess the correct key.

Imagine you want to send a secret message to a friend, but you're worried someone else might read it. Encryption is like a magic spell that turns your message into gibberish to everyone except you and your friend. First, you write your message, which is clear and understandable. Then, using a special secret code, you scramble the message into something that looks completely nonsensical, like turning "Hello" into "Xqyi!" This scrambled message is then sent to your friend. Despite anyone else possibly intercepting it, they can't make sense of it because they don't know how to unscramble it. However, your friend, who knows the secret on how to reverse the spell—or decrypt the message—uses a secret key that you both agreed on. This key transforms the gibberish back into the original, clear message. Once decrypted, your friend reads your secret, ensuring that what you wanted to say remains between just the two of you. In the digital world, this process keeps personal data safe as it travels across the internet, ensuring that only those with the correct key can read the original content.

A Brief History Of Encryption

The evolution of encryption has closely mirrored the advancements in technology. As our computational capabilities have grown, so too has the complexity of encryption methods, evolving in tandem with technological progress.

Let's take a moment to explore the origins of encryption and trace its development through the years...

The Early Days

Simple Ciphers

The concept of encryption is not new and dates back to ancient times. However, in the context of computing, encryption began with simple algorithms. In the early days of digital computing, encryption methods such as the Caesar cipher, which shifts letters by a fixed number down the alphabet, were adapted for electronic messages. These methods were relatively easy to implement and understand but offered limited security.

The Advent of Public Key Cryptography

The 1970s marked a crucial point in the history of digital encryption with the introduction of *public key cryptography*. This revolutionary approach, developed by Whitfield Diffie and Martin Hellman, introduced the concept of two keys: a public key for encrypting messages and a private key for decrypting them. This meant that people could securely communicate without needing to share a secret key in advance. RSA (Rivest–Shamir–Adleman) algorithm, introduced in 1977, became one of the first and most influential public key encryption methods, enabling secure data transmission over the internet.

RSA (Rivest–Shamir–Adleman) is one of the most significant cryptographic algorithms in the history of digital security, named after its inventors: Ron Rivest, Adi Shamir, and Leonard Adleman. Introduced in 1977, RSA was the first practical implementation of public-key cryptography, a system that transformed digital security by enabling secure, encrypted communication over insecure channels without the need for a shared secret key.

RSA operates on the principle of asymmetric cryptography, where two different keys are used: a public key for encryption and a private key for decryption. The security of RSA is based on the mathematical challenge of factoring large prime numbers, a problem that is currently infeasible to solve efficiently with existing computing technology, making RSA encryption incredibly secure.

Now, let's delve into the technical side, but don't worry—I'll keep it straightforward. I'm about to break down the brilliant concept of RSA encryption with an easy-to-follow example. This way, you'll understand how RSA encryption operates, a method still very much in use in our digital world today.

Let's break down how RSA encryption works using a simple example with small prime numbers. Remember, in the real world, RSA uses very large prime numbers to ensure security, but for the sake of understanding, we'll use smaller numbers.

Step 1: Choose Two Prime Numbers

Imagine you pick two prime numbers: 61 and 53.

Step 2: Calculate the Product

Multiply these two primes together to get a product, which we'll call n. So, *61 * 53 = 3233*. This number n is part of both the public key and the private key.

Step 3: Calculate Phi (Φ)

Calculate the totient (Φ) of *n*.

This is $(61-1)*(53-1) = 3120$. The totient function, $\Phi(n)$, for two prime numbers is just $(p-1)*(q-1)$, where p and q are your prime numbers.

Step 4: Choose Public Key Exponent

Pick a number that is relatively prime to 3120 and less than it. A common choice is 17 (it's a prime number and doesn't divide evenly into 3120).

Step 5: Calculate the Private Key

Now, find a number d such that $(d * 17) \% 3120 = 1$. A bit of math gives us $d = 2753$.

Public and Private Keys

Your public key is the pair $(n=3233, e=17)$.

Your private key is $(n=3233, d=2753)$.

Encrypting a Message

Let's say you want to encrypt the *number 123* (in real RSA, this would be the numerical representation of your message).

You use the public key to encrypt:

(message^e) % n = (123^17) % 3233 = 855

The encrypted message is *855*.

Decrypting the Message

To decrypt 855 and get back 123, you use the private key:
(encrypted_message^d) % n = (855^2753) % 3233 = 123.

And there you have it! You've successfully encrypted and decrypted a message using RSA, though with much smaller numbers than those used for real-world security.

A Little Clarification on Public and Private Keys

Imagine you have a special box where you can send and receive secret messages. This box has two different keys: a public key and a private key.

Public Key

Think of the public key like your home's mailbox address that you can share with anyone. Anyone who knows this address (or public key) can send you a letter (or encrypted message). However, once a letter is put into the mailbox, it can't be opened by just anyone because it needs a special key to unlock it—that's where the private key comes in.

Private Key

The private key is like the key to your mailbox that only you have. Even though anyone can send you letters by knowing your mailbox address, only you can open these letters because you have the unique key. In the digital world, this means that even though messages can be encrypted by anyone using your public key, only you can decrypt and read them with your private key.

So, in simple terms:

Public Key: Something you share with others so they can encrypt (secure) messages meant for you.

Private Key: Something you keep to yourself so you can decrypt (unlock) messages that were encrypted with your public key.

This asymmetry, where one key encrypts and the other decrypts, is indeed the genius of RSA and public-key cryptography as a whole. It allows for secure communication and data exchange, as only the holder of the private key can access the encrypted information, ensuring confidentiality.

Securing Data in Transit and at Rest

Encryption safeguards data in transit—data being transferred over networks, such as the internet—by ensuring that intercepted messages cannot be understood by unauthorized parties. Similarly, encryption protects data at rest—data stored on digital media—by making it inaccessible to anyone who doesn't have the decryption key. This dual protection is crucial for maintaining the confidentiality and integrity of sensitive information, from personal communications to financial transactions.

Limitations of Current Encryption Technologies

While encryption is a powerful tool for data security, it is not without its limitations.

Computational Power

The advancement in computational capabilities, including quantum computing, poses a potential threat to current encryption methods, as they might eventually be able to break complex encryption algorithms. We will explore this topic in greater detail in subsequent chapters.

Implementation Flaws

Errors in how encryption is implemented can create vulnerabilities. This includes weak algorithms, poor key management, and software bugs.

User Error

The effectiveness of encryption can be compromised by user errors, such as sharing private keys or using predictable passwords.

Regulatory Challenges

Government policies and regulations demanding access to encrypted data can undermine the strength of encryption, creating potential backdoors for unauthorized access.

Despite these challenges, encryption remains a fundamental pillar of digital security. Ongoing research and development are essential to address these limitations and ensure that encryption technologies can withstand new threats in our ever-evolving digital landscape.

Data Integrity and Verification

Before concluding our discussion on data security, there's an essential aspect I'd like to cover: verifying data integrity. In simpler terms, how can we be certain that a dataset remains authentic and unchanged, unaffected by alteration or damage during storage? Data integrity is the assurance of data's accuracy and consistency throughout its lifecycle. It guarantees that information remains dependable, capable of being retrieved exactly as it was initially stored or transmitted. The importance of data integrity spans several crucial areas, including operational efficiency, adherence to regulations, and the safeguarding of security. Compromises to data integrity may arise from inadvertent errors, such as hardware failures or software glitches, as well as intentional interference, including cyber attacks or data manipulation.

The Role of Checksums in Verifying Data

A checksum is a simple yet powerful mechanism designed to verify the integrity of data. It works by calculating a small-sized datum from a block of digital data for the purpose of detecting errors that may have been introduced during its storage or transmission. The basic idea is to sum up the total value of all the bits in the data packet or file, then store this sum along with the data. When the data is later retrieved or received, the checksum calculation is performed again on the data. If the new checksum matches the original one, it's highly likely that the data has not been altered and is therefore correct.

How Checksum Works

Step 1: Calculation

When data is created or before it's transmitted, a checksum value is calculated. This involves running the data through a specific algorithm, which produces a short, fixed-length value— the checksum.

Step 2: Storage or Transmission

The checksum is then either stored with the data (if the data is being saved) or sent along with the data (if the data is being transmitted).

Step 3: Verification

Upon retrieval or reception, the data undergoes the same checksum calculation. The newly calculated checksum is compared to the original checksum.

Step 4: Integrity Assessment

If the two checksums match, the data is considered intact and unaltered. If they don't match, it indicates that the data may have been corrupted or tampered with, prompting further investigation or retransmission.

Checksum algorithms vary in complexity and reliability, from simple ones like the Longitudinal Parity Check and Cyclic Redundancy Check (CRC) to more sophisticated hash functions like MD5 and SHA-256. While not foolproof, especially against sophisticated tampering, checksums provide a first line of defense in maintaining data integrity.

Digital Signatures

Imagine you're sending a letter to a friend, and you want to make sure they know it's genuinely from you and that no one has tampered with it along the way. A digital signature serves a similar purpose in the digital world. It's like an electronic stamp you put on digital documents or messages to prove that you are the sender and that the message hasn't been changed after you signed it.

Here's how digital signatures work in simplified terms...

Step 1: Creating the Signature

First, when you have a document or message you want to send securely, you create a "hash" of it. Think of this hash as a unique, fixed-size summary or fingerprint of your message. Even a small change in the message would result in a completely different hash.

Then, you encrypt this hash with your private key, which is a secret code that only you have. This encrypted hash is your digital signature.

Step 2: Attaching the Signature

You then send the original message along with the digital signature to your friend.

Step 3: Verifying the Signature

- > Your friend receives both the message and the digital signature. To make sure the message is truly from you and hasn't been tampered with, they decrypt the digital signature using your public key (which is like the public address everyone knows is yours).

- > Decrypting the signature with your public key reveals the original hash value you created.

- > Your friend then creates a hash of the received message (the same way you did) and compares it to the hash that was decrypted from the digital signature.

- > If the two hashes match, it means the message hasn't changed since you signed it, and it's genuinely from you.

The beauty of digital signatures lies in their ability to ensure the authenticity and integrity of digital communications. The use of a private key for signing and a public key for verification means that anyone can verify the sender's identity and that the message is intact, but only the sender can create the signature. This system is widely used for secure email, online transactions, and document verification, providing a high level of trust and security in the digital world.

Hash Functions

Imagine you have a magical blender that transforms any food you put into it into a smoothie of a specific color. No matter how much or what combination of food you add, the resulting smoothie's color is unique to that exact mixture. If you change even a tiny bit of the ingredients, the color changes dramatically. In the digital world, hash functions are like this magical blender for data.

What are Hash Functions?

A hash function is a mathematical process that takes input data of any size (like a long email, a document, or a video) and produces a fixed-size string of characters, which is typically a sequence of numbers and letters. This output is known as a hash value or hash code.

How Hash Functions Work

-> Input of Any Size

You can feed any amount of data into a hash function, large or small.

- > Fixed-Size Output

No matter the size of the input, the hash function produces a hash of a specific, fixed length.

- > Unique Output

For any unique input, the hash function produces a unique output. If you input the same data twice, you get the exact same hash. But if the input changes even slightly, the resulting hash is entirely different.

How Hash Functions are Used

- > Verifying Data Integrity

Hash functions help check if data has been altered. When you download a file, you can hash it and compare the result to the original hash provided by the source. If they match, the file hasn't changed. If they differ, the file might have been tampered with.

- > Storing Passwords

Instead of storing actual passwords, websites store the hash values of those passwords. When you log in, the website hashes the password you enter and compares it to the stored hash. This way, even if someone gains access to the database, they only see hash values, not real passwords.

- > Blockchain and Cryptography

In cryptocurrencies like Bitcoin, hash functions secure transactions and ensure the integrity of the blockchain. Each block is identified by a hash, linking it securely to the previous block and making tampering practically impossible.

Hash functions are a fundamental pillar of computer security and data integrity, making them indispensable in our increasingly digital world. They provide a way to securely manage data, protect sensitive information, and verify that digital assets remain unchanged and authentic.

Data Redundancy and Replication

Another vital method for maintaining data integrity and ensuring accessibility involves the concepts of data replication and redundancy. This technique is widely employed to protect data against hardware corruption or damage. Simply put, when data is saved, it's as if you're storing it across several independent hardware systems rather than just one. Consequently, if one of the storage units becomes corrupted, there are additional copies elsewhere that can be utilized to recover the lost data. Now, let's dive into the meanings of data replication and data redundancy...

Data Redundancy involves storing additional copies of data within the same or across different storage devices. This method acts as a fail-safe, ensuring that if one copy becomes corrupted or inaccessible, other copies remain unaffected, thereby preserving the integrity and availability of data.

Replication, on the other hand, entails copying data from one location to another, typically in real-time or at scheduled intervals. This ensures that an up-to-date copy of the data is always available in a secondary location, ready to be accessed in case the primary source fails.

The Ingenious Technology of RAID

RAID, standing for Redundant Array of Independent Disks, represents a remarkable technology designed to enhance data performance and ensure data redundancy. To put it simply, imagine RAID as a team of workers, each assigned a specific task. Together, they accomplish tasks more swiftly and guarantee that no critical work is compromised, even if one member is absent. There are various types of RAID, each tailored to serve distinct purposes, ranging from highly redundant to highly efficient data storage solutions.

RAID (Redundant Array of Independent Disks) combines multiple hard drives into one system to protect data and improve performance. Here's a look at the different types of RAID and how they work...

RAID 0 – Striping

Increases performance

RAID 0 involves dividing data into blocks, with each block then being written to a different disk drive.

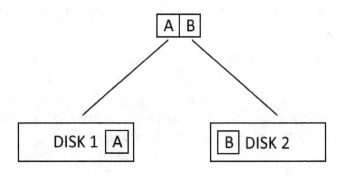

RAID 0 – Striping

This setup enhances performance by enabling multiple disks to be read from or written to at the same time. However, it lacks redundancy; therefore, if any single disk in the array fails, all the data stored across the disks is compromised and lost.

RAID 1 - Mirroring

Provides redundancy

RAID 1 ensures data safety by mirroring; that is, data written to one disk is concurrently copied to another disk, creating an identical mirror.

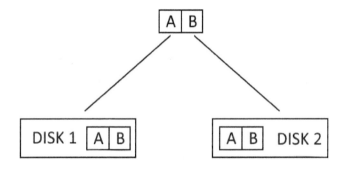

RAID 1 - Mirroring

This redundancy means that should one disk fail, the mirrored disk serves as a backup from which data can be retrieved. The primary limitation of this configuration is its efficiency in storage utilization: only 50% of the total disk capacity is available for actual storage, with the remaining half dedicated to the mirrored copy.

RAID 5 – Striping with Parity

Balances performance with redundancy

RAID 5 distributes both data and parity information (which is essential for data recovery in the event of a disk failure) across all the disks in the array.

Parity serves as an ingenious method for verifying whether data has been altered, either inadvertently during storage or in the course of transmission. This is achieved by tallying the bits (the 1s and 0s that make up digital data) and appending a 0 or a 1 to the end, ensuring that the total count of 1s is either even or odd, based on the parity scheme used.

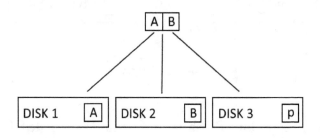

RAID 5 – Striping with Parity

RAID 5 balances performance with redundancy. Allows recovery from a single disk failure. The drawback however is that it is slightly slower write performance due to the calculation of parity information.

RAID 6 – Striping with Double Parity

Greater Fault Tolerance

RAID 6 employs a technique known as striping with double parity, functioning similarly to RAID 5 but enhanced by an additional layer of parity data.

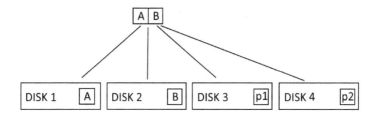

RAID 6 – Striping with Double Parity

This extra parity enables RAID 6 to tolerate the failure of two disks, offering superior fault tolerance compared to RAID 5. However, this level of protection comes with certain requirements and limitations: it necessitates the use of at least four disks and results in a reduction of available storage space due to the allocation of extra space for the added parity data.

RAID 10 (1+0) – Mirroring and Striping

High Performance and High Redundancy

RAID 10 (1+0) merges the techniques of mirroring and striping, integrating the strengths of RAID 1 and RAID 0.

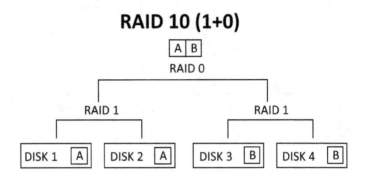

RAID 10 – Mirroring and Striping

In this setup, data is first duplicated across pairs of disks for redundancy, and then these mirrored pairs are striped together, enhancing performance. This configuration delivers the dual benefits of high performance and robust redundancy. However, it necessitates the use of at least four disks, and because of the mirroring, only half of the overall disk capacity is available for actual data storage.

PART III

Data

Storage

The Hardware

Having gained an understanding of what data is, along with its structure and security within the digital world, our attention now shifts to the physical vessels of data storage. This chapter will begin an exploration through the history of data storage units, tracing their development over time. We will explore the various types of storage devices that are currently in use, encompassing everything from conventional options to cutting-edge solutions designed for long-term preservation, capable of safeguarding data for up to a thousand years. Additionally, we will examine the challenges associated with maintaining data integrity over extended periods, highlighting the innovations and obstacles in the quest for enduring data storage.

I still vividly recall my very first computer, the Sinclair Spectrum. Unlike the electronic devices of today, my

Spectrum didn't come with any built-in storage. To work on anything, you had to hook up your home stereo system to the computer and load programs from a cassette tape. The Spectrum could only manage to load a maximum of 32,000 bytes of data—a figure that's at least a million times smaller than the capacity of modern computers. The process of loading just 32,000 bytes could take upwards of 15 minutes, a pace that's now millions of times slower compared to current technologies. I remember those moments of anticipation, waiting 10-15 minutes, only for the computer to report a failure to load the program, prompting me to start the process all over again in hopes of success. From these early days, data storage has undergone a remarkable transformation. Let's dive into the history and evolution of storage units, comparing their capacities to the technology we use today.

A Brief History of Storage Devices

The history of digital storage begins in the early days, where data storage was rudimentary and manual, and evolves to the sophisticated, high-capacity devices we rely on today.

The Era of Punch Cards

The story of digital storage starts in the late 19th century with punch cards. Initially used for controlling looms in the textile industry, punch cards were later adapted by Herman Hollerith to process data for the 1890 U.S. Census. These cards stored information in the form of holes punched in specific locations, representing data that machines could read and tabulate. Despite their simplicity, punch cards were a significant step forward in data processing and storage.

Magnetic Tapes and Floppy Disks

Magnetic Tapes

In the 1950s and 1960s, magnetic tapes were the primary medium for data storage in commercial settings, especially for large-scale computers. Their storage capacity ranged from a few kilobytes to several megabytes per reel, catering to the data storage needs of the time, which were relatively low compared to today's standards.

Floppy Disks

Introduced in the late 1960s, floppy disks became a game-changer for personal computing in the 1970s and 1980s. The first 8-inch floppies could store about 80 kilobytes, but by the time 3.5-inch disks became standard, they boasted capacities up to 1.44 megabytes—a significant amount for personal users at the time.

The Optical Revolution

CDs, DVDs, and Blu-ray Discs

CDs (Compact Discs)

Launched in the 1980s for audio and later adapted for data (CD-ROM), CDs could store approximately 700 megabytes of data. This marked a substantial increase in storage capacity for personal users and began to find its way into commercial applications for software distribution.

DVDs

Building on CD technology, DVDs emerged in the 1990s with a storage capacity of 4.7 gigabytes for a single-layer, single-sided disc, and up to 17.08 GB for double-layer, double-sided discs. DVDs offered a balance of capacity and accessibility for both commercial and personal use, facilitating larger software applications and video content.

Blu-ray Discs

As high-definition content became standard, Blu-ray discs entered the market in the mid-2000s with a single-layer capacity of 25 gigabytes and a dual-layer capacity of 50 gigabytes, providing ample space for high-definition video and large-scale data storage needs.

The Era of Hard Drives and Solid-State Drives

Hard Disk Drives (HDDs)

The evolution of HDDs has seen capacities grow exponentially, from megabytes in the early models to several terabytes in today's drives. Commercial enterprises rely heavily on large-capacity HDDs for data centers and cloud storage, while personal computers commonly use drives ranging from 500 gigabytes to several terabytes.

Solid-State Drives (SSDs)

SSDs represent the latest advancement in storage technology, with no moving parts and faster data access speeds than HDDs. While more expensive per gigabyte than HDDs, SSDs offer superior performance and reliability. Personal computers increasingly use SSDs, with common sizes ranging from 256 gigabytes to 2 terabytes, while commercial applications leverage larger SSDs for high-demand environments.

Commercial vs. Personal Storage Systems

Commercial data storage units and personal use disks differ primarily in scale, performance, durability, and the technologies they employ to meet the distinct needs of businesses versus individual consumers.

Personal Use Disks

Personal use disks, encompassing external hard drives, USB flash drives, and personal cloud storage, are tailored to meet individual storage needs with capacities that range from a few gigabytes (GB) to a few terabytes (TB). These devices, while offering fast data transfer rates thanks to technologies like USB 3.0 and Thunderbolt, generally focus less on high performance compared to their commercial counterparts. Designed for lighter usage, personal storage disks may not exhibit the same degree of durability or redundancy as seen in commercial units, yet they remain dependable for daily tasks and data backups. Emphasizing simplicity and convenience, consumer storage solutions are crafted for ease of use, featuring plug-and-play capabilities that facilitate the straightforward storage of photos, videos, documents, and other personal content without the need for complex setup procedures. Moreover, personal use disks are significantly more cost-effective, mirroring their simpler construction and the reduced demands placed upon them relative to more sophisticated commercial storage options.

Commercial Data Storage Units

Commercial data storage units are engineered to manage vast quantities of data, ranging from terabytes (TB) to petabytes (PB) and even more, catering to the comprehensive data storage and access requirements of whole organizations. This includes accommodating large databases, customer information, and transaction records among others. The performance of these storage solutions is pivotal, with speed and reliability being paramount to support business operations and ensure customer satisfaction. Advanced technologies are employed within enterprise storage systems to facilitate high data transfer rates and swift access to information. In terms of durability and reliability, commercial storage systems are constructed to endure continuous operation and heavier workloads, incorporating redundant power supplies, data paths, and networking connections to maintain data availability, even when hardware failures occur. Additionally, enterprise storage solutions are equipped with elaborate data management features such as automated backups, disaster recovery, data deduplication, and hierarchical storage management, aiding businesses in optimizing storage efficiency and safeguarding against data loss. Due to their sophisticated features and superior capacities, commercial storage units come with a higher price tag compared to personal storage devices, an investment that underscores their vital contribution to business continuity and data protection.

What is a Data Center?

A commercial data center is a facility used to house computer systems and associated components, such as telecommunications and storage systems.

Data Center

It is designed to provide a reliable, secure, and efficient environment for businesses to process and store critical data. The structure of a commercial data center involves various components and design considerations to ensure optimal performance, security, and scalability. Here's a general overview of how a commercial data center is structured...

Physical Infrastructure

The physical infrastructure of a data center is meticulously designed to encompass multiple layers of security checkpoints, with features such as mantraps and biometric access controls fortifying the perimeter. Within its walls, the center is methodically divided into distinct zones dedicated to servers, storage, networking equipment, and support infrastructure. To maintain an environment conducive to optimal performance and hardware longevity, efficient cooling systems—including in-row cooling, raised floors for air distribution, and liquid cooling—are implemented. Additionally, the data center's reliance on a redundant power supply, characterized by UPS (Uninterruptible Power Supplies) systems and backup generators, guarantees uninterrupted operation even during power outages. Furthermore, the installation of advanced fire detection and suppression systems plays a critical role in safeguarding personnel and equipment, effectively preventing and mitigating fire-related damages.

Networking Infrastructure

The networking infrastructure is built on a foundation of multiple high-speed, redundant internet connections to guarantee uninterrupted service, utilizing both major ISPs and direct peering arrangements for robust connectivity. It is equipped with a comprehensive array of network hardware, including routers, switches, firewalls, and load balancers. These components are crucial for managing traffic flow, safeguarding against cyber threats, and evenly distributing network loads, thereby ensuring consistent service availability and speed.

Server and Storage Infrastructure

The server and storage infrastructure is anchored by racks of servers that deliver the necessary computational power for processing and running applications, often dedicated to specific tasks like web hosting, database management, and application hosting. Complementing these servers, the infrastructure includes SAN (Storage Area Network) and NAS (Network Attached Storage) systems, providing scalable storage solutions that enable data redundancy, ensure high availability, and facilitate efficient data management.

Security and Monitoring

The security and monitoring of data centers are comprehensive, encompassing physical security measures such as electronic access controls, the deployment of security personnel, the use of surveillance cameras, and the implementation of intrusion detection systems to prevent unauthorized access. In terms of cybersecurity, the centers are fortified with advanced measures that include firewalls, intrusion detection and prevention systems (IDPS), and the conduct of regular security audits to guard against online threats. Additionally, environmental monitoring is conducted through sensors and monitoring systems that track humidity, temperature, and power usage effectiveness (PUE), ensuring that the data center's environment remains within the designated operational parameters.

Scalability and Management

Data centers often embrace a modular design to facilitate straightforward expansion in response to increasing demand, ensuring such growth does not interrupt ongoing operations. To complement this architectural approach, data center infrastructure management (DCIM) software plays a crucial role by offering a comprehensive overview of a data center's performance. This software aids in the efficient management of resources and operations, contributing to the overall efficacy of the facility. In essence, the structure of a commercial data center is carefully devised and upheld to guarantee high availability, security, and operational efficiency, thereby meeting the expanding data requirements of businesses and reducing risks tied to data management and processing.

The Role of Data Centers in the Digital Age

In our increasingly interconnected world, the importance of data centers cannot be overstated. Every aspect of our daily lives involves interaction with vast quantities of data. Whether we're working, relaxing at home, traveling, or even sleeping, we're constantly generating and storing data on the "Cloud." This term refers to the global network of data centers that safeguard our information and make it accessible over the internet. Below, we explore the diverse uses of these colossal data centers that play a pivotal role in our digital existence...

Hosting Websites and Applications

Data centers provide the infrastructure necessary to host websites and applications, ensuring they are accessible to users around the world 24/7. This includes everything from corporate websites and e-commerce platforms to mobile apps and online services.

Cloud Computing Services

They are central to the provision of cloud computing services, including software as a service (SaaS), platform as a service (PaaS), and infrastructure as a service (IaaS). Businesses rely on these services for everything from email and collaboration tools to database management and development environments.

Data Storage and Management

Commercial data centers offer secure and scalable solutions for data storage and management, accommodating the vast amounts of data generated by businesses today. This includes customer data, transaction records, business analytics, and more.

Business Continuity and Disaster Recovery

By housing backup servers and data storage, data centers play a crucial role in business continuity and disaster recovery plans. They ensure that, in the event of a disaster or data loss incident, businesses can quickly recover their data and resume operations with minimal downtime.

Content Delivery Networks (CDNs)

Data centers are key nodes in content delivery networks (CDNs), which distribute web content and media to users worldwide. CDNs cache content at strategic locations to improve website performance and reduce latency for end-users.

Networking and Internet Connectivity

Offering high-speed, reliable internet connectivity and networking services, data centers enable businesses to connect with customers, partners, and employees efficiently. They facilitate the exchange of data and support telecommunication services.

Big Data Analytics and Artificial Intelligence

The computational power and storage capabilities of data centers support big data analytics and artificial intelligence (AI) applications. Businesses leverage these technologies for insights into customer behavior, operational efficiency, and strategic decision-making.

Support for IoT (Internet of Things)

As IoT devices proliferate, generating vast quantities of data, data centers provide the infrastructure needed to process, analyze, and store IoT-generated data, supporting applications ranging from smart homes to industrial automation.

NAS Drives for Personal Use

A Gateway to Efficient Data Management

In the domain of personal data storage, Network Attached Storage (NAS) drives emerge as a powerful solution that transcends the limitations of traditional hard drives. NAS drives offer a seamless blend of accessibility, flexibility, and security, catering to the evolving needs of digital lifestyles.

Integrating a Network Attached Storage (NAS) system into your home or office essentially equips you with a mini data center, tailored for your personal or small business needs. Let's explore the advantages of incorporating NAS drives for your personal use or in your business.

Centralized Storage Hub

At its core, a NAS drive acts as a centralized storage hub, allowing users to store all their digital content in one place. From precious family photos and expansive media libraries to important documents and backups, a NAS drive consolidates data, eliminating the need for multiple external hard drives cluttering your workspace.

Anywhere Access

One of the hallmark features of NAS drives is their ability to provide access to your data from anywhere with an internet connection. Unlike traditional storage solutions that tether your data to a specific device, NAS drives offer the convenience of remote access. Whether you're at home, in the office, or halfway across the globe, your data is just a few clicks away, ensuring that important files are always within reach.

Streamlining Media and Entertainment

NAS drives excel in streamlining media consumption and entertainment. With the capability to serve as a home media server, NAS drives support streaming of videos, music, and photos to various devices such as smart TVs, game consoles, and smartphones. This functionality transforms your NAS into a personal Netflix or Spotify, offering a tailored entertainment experience without the confines of third-party platforms.

Data Protection and Redundancy

Data protection is crucial in today's digital age, and NAS drives offer robust solutions to safeguard against data loss. Many NAS systems come with built-in redundancy features, such as RAID configurations, which store duplicates of your data across multiple disks. In the event of a disk failure, your data remains intact and recoverable, providing peace of mind and ensuring the longevity of your digital archives.

Simplified Collaboration and Sharing

NAS drives facilitate simplified collaboration and sharing, making it easy to share files and collaborate on projects with family members or colleagues. Permissions and access controls allow you to manage who can view or edit certain files, streamlining the collaborative process and enhancing productivity.

Pros and Cons of NAS vs. Cloud Storage

In the vast landscape of digital storage solutions, the choice between Network Attached Storage (NAS) and cloud storage presents a critical decision for individuals and businesses alike. Each option carries its distinct set of advantages and challenges, tailored to different needs and operational frameworks.

Network Attached Storage stands out for its local control, offering users complete authority over their data management and security protocols. This control translates into superior performance, particularly for large file transfers within the same network, making NAS an attractive proposition for those with intensive data access needs. The investment in NAS is primarily upfront, encompassing the cost of the device and the drives, without recurring fees for basic usage. Furthermore, NAS systems boast expandability, allowing users to seamlessly increase storage capacity as their needs evolve.

However, the initial cost for a comprehensive NAS setup can be substantial, reflecting both the hardware and necessary drives. Ownership also implies the responsibility for maintenance and management, including tackling hardware issues, performing software updates, and executing data backups. Additionally, NAS systems depend on consistent power and internet connectivity to function, with accessibility potentially compromised during outages or network disruptions.

Conversely, cloud storage excels in accessibility, enabling users to access data from any location with internet connectivity. This flexibility supports remote work and simplifies data sharing across geographies. Cloud services are inherently scalable, allowing for effortless adjustments in storage capacity to match current demands.

Maintenance responsibilities fall to the service provider, who ensures the system's reliability, security, and currency. Advanced security features and redundancy across multiple locations further enhance data protection and recovery prospects.

Yet, cloud storage is not without its drawbacks. The model typically involves ongoing costs based on storage volume and additional services, which can accumulate over time. Dependence on a stable internet connection for access means that slow or unreliable connections can impede effective data usage. Concerns also arise regarding data privacy and control when storing sensitive information with third-party providers, compounded by the complexities of compliance and legal standards in different jurisdictions. Additionally, the performance of cloud storage, particularly for operations involving large files, may lag behind local NAS solutions due to inherent bandwidth limitations.

Choosing between NAS and cloud storage, thus, hinges on weighing these factors against individual priorities and requirements.

For those valuing direct control, performance, and a one-time investment, NAS emerges as a compelling choice. In contrast, cloud storage offers unmatched accessibility, ease of scalability, and freedom from maintenance burdens, suitable for those comfortable with subscription costs and navigating data privacy in a cloud environment. A hybrid approach, combining the strengths of NAS for handling sensitive or heavily accessed data with the scalability and remote access benefits of cloud storage, often represents a balanced, versatile solution in the modern digital storage paradigm.

~~~~~~

With this, we reach the end of our chapter. I hope that by now, you've gained a thorough understanding of the hardware aspect of data management, spanning both commercial and personal contexts. In the upcoming chapter, we will pivot our attention towards the utilization of data, exploring the myriad ways in which it is applied and leverages value.

PART IV

Data Analysis & Utilization

Connecting The Dots

Introduction to Data Analysis and Utilization

With vast amounts of data at our fingertips, the subsequent step is unraveling how this collected data is wielded across various systems to sculpt the world around us. Data analysis is a daily influence on each of our lives, subtly shaping decisions, policies, and innovations in ways both seen and unseen.

The modern world thrives on information, and data analysis stands at the forefront of converting raw data into actionable insights. This transformative process touches every corner of society, empowering businesses to optimize operations, enabling governments to craft informed policies, and assisting individuals in making choices that align with their goals and needs.

At the heart of data analysis lies the ability to sift through datasets, identifying patterns, trends, and correlations that might not be immediately apparent. This analytical prowess turns the abstract into the tangible, offering invaluable insights that drive decision-making processes. For businesses, it means pinpointing customer needs, streamlining production, and staying ahead in competitive markets. For governments, it translates into understanding societal trends, allocating resources efficiently, and enhancing public services. For individuals, data analysis can personalize experiences, from tailored recommendations on streaming platforms to customized health and fitness plans.

Data analysis encompasses a variety of methods and approaches, each suited to unpacking different layers of information hidden within datasets. At its foundation, data can be categorized into two primary types: *quantitative and qualitative*. Quantitative data is numerical, providing a measurable insight into problems, such as sales figures or website traffic statistics. Qualitative data, though not numerical, offers depth and context through observations, interviews, and open-ended responses, painting a detailed picture of user experiences and preferences.

The general process of data analysis follows a structured path from collection to conclusion. It begins with defining the questions we seek to answer, followed by collecting relevant data. This data then undergoes cleaning and preprocessing to remove inaccuracies or irrelevant information. The next phase involves exploratory data analysis, where patterns and trends are identified, leading to more in-depth analysis using statistical methods or machine learning algorithms. The final step is interpreting these results, drawing conclusions, and making informed decisions or recommendations.

As we dive deeper into this chapter, we will uncover the tools, techniques, and real-world applications of data analysis and utilization. The journey from data to insight is complex, yet immensely rewarding, offering a lens through which we can better understand and navigate the world around us. Data analysis is the bridge between the vast seas of information we've learned to collect and store, and the actionable insights that propel us forward into a more informed future.

The Process of Data Analysis

Data analysis is a meticulous process that transforms raw data into valuable insights, guiding decisions across various domains. This journey from data to decision encompasses several critical steps, each pivotal in ensuring the integrity and usefulness of the final analysis. Let's navigate through these steps, from the initial collection to the final interpretation, and explore the diverse methods and techniques that breathe life into data.

Step 1: Data Collection

The foundation of any data analysis is the collection phase. This initial step involves gathering relevant data from multiple sources, which could range from internal databases and surveys to external datasets and online repositories. The objective here is to amass data that is accurate and relevant to the specific questions or hypotheses being investigated.

 The business distributes a survey to its customers asking various questions about their satisfaction levels, preferences, and suggestions for improvement. Responses are collected through an online platform, resulting in a dataset comprising various types of responses, including numerical ratings and textual feedback.

Step 2: Data Cleaning and Preparation

Once collected, data rarely arrives in a ready-to-analyze state. It often contains errors, inconsistencies, or missing values that could skew the analysis. The cleaning phase addresses these issues, involving tasks such as removing duplicates, handling missing values, and correcting errors. Preparation further involves transforming and organizing data into a suitable format for analysis, ensuring it is primed for the next steps.

 Next comes data preparation, which involves cleaning and organizing the collected data. This step might include:

- Removing incomplete responses: Surveys that weren't fully completed are excluded to ensure data quality.

- Standardizing responses: Ensuring all data is in a consistent format, such as converting all text to lowercase for textual analysis or ensuring date formats are uniform.

- Categorizing textual feedback: Using natural language processing (NLP) techniques to categorize open-ended responses into themes like "pricing," "product quality," and "customer service."

Step 3: Data Exploration

Before diving deep into complex analysis, data exploration offers a preliminary look at the data. This step involves summarizing the main characteristics of the dataset through descriptive statistics and visualizations. It helps identify patterns, outliers, and anomalies, which can inform more focused analyses later on.

Step 4: Data Analysis

With the data cleaned and explored, the core phase of analysis begins. This stage applies various statistical, machine learning, and data mining methods to uncover deeper insights and relationships within the data.

- **Statistical Analysis** involves applying statistical techniques to test hypotheses or uncover relationships between variables. Common methods include regression analysis, hypothesis testing, and variance analysis.

- Machine Learning leverages algorithms that enable computers to learn from and make predictions or decisions based on data. It includes supervised learning for predictive modeling and unsupervised learning for finding patterns or groupings in data.

To learn more about AI and Machine Learning, please check out my AI Simplified book.

AI Simplified

- Data Mining is the process of exploring large datasets to find meaningful patterns, relationships, or trends. It employs techniques like clustering, association analysis, and anomaly detection to discover insights that aren't readily apparent.

EXAMPLE

With the data cleaned and organized, we move to analysis. One technique we might employ is clustering, particularly for the numerical satisfaction ratings. Clustering involves grouping data points (in this case, customer ratings) that are similar to each other. For instance:

- *K-means clustering:* We apply K-means clustering to segment customers into groups based on their satisfaction ratings. This could reveal clusters such as "highly satisfied," "moderately satisfied," and "dissatisfied" customers.

Step 5: Data Interpretation

The culmination of the data analysis process is interpretation. This step involves making sense of the results, drawing conclusions, and often translating these findings into actionable recommendations. Interpretation requires not just technical acumen but also domain knowledge and critical thinking to ensure the insights are valid, relevant, and capable of driving informed decisions or actions.

Finally, we reach the stage of data visualization, where we turn our analyzed data into interpretable and engaging visuals:

- Bar charts or pie charts: These could represent the distribution of customer satisfaction levels, showing at a glance how many customers fall into each satisfaction category.

- Word clouds: For textual feedback, a word cloud could visualize the most frequently mentioned words in customer comments, highlighting key areas of concern or praise.

- Cluster diagrams: To visualize the results of the K-means clustering, a scatter plot with color-coded points could show how customers are grouped according to their satisfaction levels, providing clear visual insight into customer sentiment.

Step 6: Communication

Although not always listed as a formal step in the data analysis process, effectively communicating the findings is crucial. This can involve creating reports, dashboards, or presentations that convey the insights in a clear and compelling manner to stakeholders or decision-makers.

The process of data analysis is both an art and a science, requiring a blend of precision in execution and creativity in interpretation. By adhering to these steps and applying the appropriate methods and techniques, analysts can unlock the true potential of data, turning it into a powerful tool for understanding and influencing the world around us. Whether through statistical analysis, machine learning, or data mining, the goal remains the same: to derive insights that inform better decisions, drive innovation, and create value.

Understanding Big Data

Before we dive deeper into our exploration of Data Analytics, let's take a moment to clarify a concept that frequently emerges in the world of Data Analytics.

In today's digital age, the term *"Big Data"* frequently surfaces across various industries, highlighting a significant shift in how we collect, analyze, and leverage information. At its core, Big Data refers to extraordinarily large datasets that traditional data processing software can't handle efficiently. But what makes Big Data truly fascinating isn't just its volume; it's about the insights and value that can be extracted from this vast sea of information. Let's break down Big Data into simpler terms to understand its essence and impact.

The Three Vs of Big Data

Big Data is often characterized by three key attributes, known as the Three Vs.

Volume

This refers to the sheer amount of data being generated every second from multiple sources like social media, business transactions, online interactions, and IoT (Internet of Things) devices. We're talking about data accumulating in orders of magnitude that range from terabytes to petabytes and beyond.

Velocity

The speed at which new data is generated and collected is staggering. With real-time processing, Big Data flows in at an unprecedented rate, requiring prompt analysis to capture timely insights.

Variety

Big Data comes in various formats - from structured, numeric data in traditional databases to unstructured text documents, emails, videos, audios, and financial transactions. This diversity presents unique challenges in capturing, storing, and analyzing the data.

The Value of Big Data

The true potential of Big Data lies in its utilization. By applying advanced analytics techniques and technologies, businesses and organizations can uncover deep insights that were previously inaccessible. These insights can lead to more informed decision-making, predicting trends, enhancing customer experiences, optimizing operations, and discovering new revenue opportunities.

Simplifying Big Data

Imagine you're trying to understand the behavior of a vast forest from a handful of leaves. Traditional data analysis might look at each leaf individually, but with Big Data analytics, you can observe the entire forest, understanding patterns, seasons, and the ecosystem as a whole. It's like having a bird's eye view of data, where every interaction, transaction, and social media post contributes to a larger picture.

Big Data technologies, such as Hadoop and Spark, act as the tools that allow us to manage this forest of information. They help store, process, and analyze the data, no matter how vast or fast it grows.

By harnessing the power of Big Data, we can transform overwhelming volumes of information into actionable insights, driving innovation and efficiency across all sectors of society. In essence, Big Data gives us the keys to understand the complex, fast-moving digital world around us, making sense of the information overload in our quest for progress.

Tools and Technologies for Data Analysis

In the quest to unlock the stories hidden within data, analysts rely on an arsenal of tools and technologies. These instruments range from the basic, like spreadsheets, to the highly sophisticated, such as programming languages and big data platforms. Each tool serves a unique purpose, catering to different aspects of the data analysis process and varying levels of complexity.

Spreadsheets

The Foundation of Data Analysis

At the most fundamental level, spreadsheet applications like Microsoft Excel and Google Sheets offer a versatile and user-friendly interface for data analysis. These tools are ideal for managing smaller datasets, performing basic calculations, creating simple models, and visualizing data through charts and graphs. Their accessibility makes them an indispensable part of an analyst's toolkit, suitable for a wide range of tasks from budgeting and planning to preliminary data exploration.

Statistical and Data Analysis Software

For more complex analyses, especially those involving large datasets or advanced statistical methods, specialized software comes into play.

"R"

An open-source programming language and environment, R is particularly strong in statistical computing and graphics. It boasts a vast repository of packages for various types of data analysis, making it a favorite among statisticians and researchers.

Python

Known for its simplicity and readability, Python has become a staple in data science thanks to libraries like *NumPy, pandas, SciPy, and scikit-learn.* These tools turn Python into a powerful platform for data manipulation, statistical analysis, and machine learning.

SAS (Statistical Analysis System)

A suite of software tools developed by SAS Institute, SAS is widely used in commercial and academic settings for data management, advanced analytics, multivariate analysis, and predictive modeling. Its robustness and extensive support make it a trusted choice for enterprise-level data analysis.

Big Data Technologies and Platforms

The explosion of data in the digital age necessitated the development of technologies capable of handling data at an unprecedented scale—enter Big Data technologies.

Hadoop

An open-source framework designed to store and process large volumes of data across clusters of computers using simple programming models. Hadoop's ecosystem, including HDFS for storage and MapReduce for processing, makes it possible to handle petabytes of data.

Spark

Another open-source project, Apache Spark, is known for its speed and ease of use in big data processing. Spark can perform batch processing (like Hadoop) but is particularly noted for its ability to handle stream processing and real-time data analysis.

Cloud-based Analytics Platforms

Providers like Amazon Web Services (AWS), Microsoft Azure, and Google Cloud Platform offer a range of services for data storage, processing, and analysis. These platforms provide the scalability and flexibility to manage large datasets without the need for physical infrastructure.

Utilization of Analyzed Data

In the vast and complex world of data analytics, the true power of data comes not just from its analysis but from how the insights gleaned are applied across various sectors. The utilization of analyzed data has become a cornerstone for innovation, efficiency, and growth in numerous fields. From enhancing customer experiences to driving policy decisions, the impact of data analysis is profound and far-reaching.

Transforming Industries with Data

Transforming Industries with Data has become a pivotal force in reshaping how businesses and organizations operate across various sectors. In retail and service, data analysis equips companies with a deeper understanding of customer behavior, preferences, and feedback, enabling them to customize their offerings. This tailored approach not only meets customer needs more precisely but also fosters increased satisfaction and loyalty.

In manufacturing, logistics, and supply chain management, the benefits of data analytics are manifold. Businesses are able to pinpoint inefficiencies, forecast maintenance requirements, and streamline resource allocation. Such optimizations lead to significant cost savings and a boost in productivity, marking a leap forward in operational excellence.

The influence of data extends to the corridors of governance and public administration, where data analytics serves as a compass for policy-making and the enhancement of public services. Through the lens of analyzed data, governments can gain insights into societal trends, public health metrics, and educational needs, laying the groundwork for policies that are both informed and impactful.

Furthermore, the scientific community harnesses data analysis as a catalyst for research and discovery. The meticulous examination of data gathered from experiments, simulations, and natural observations opens up new vistas of knowledge in diverse fields such as genomics, environmental science, and physics.

The Role of Data Visualization

Data visualization emerges as a crucial element in data analysis, serving as a bridge between complex data sets and their practical application. It elevates the process of data analysis by not only uncovering hidden patterns and insights but also ensuring these findings are communicated effectively for their intended use. This transformation of intricate data into intuitive, visual formats allows stakeholders of varying levels of data science expertise to easily comprehend the information presented.

By employing charts, graphs, and infographics, data visualization significantly enhances comprehension, allowing statistical nuances and trends to be quickly understood at a glance. Such visual representations are adept at illuminating relationships and patterns within the data that may remain obscured in traditional, numerical formats.

Moreover, data visualization is instrumental in facilitating communication. It empowers analysts to convey their findings in a manner that is both compelling and easy to digest, which is paramount in the context of decision-making processes. The ability to present data-driven insights clearly and concisely can substantially impact the strategic direction of businesses and initiatives.

Perhaps most importantly, data visualization plays a vital role in driving action. By rendering data both accessible and comprehensible, visual representations encourage stakeholders to act upon the insights provided. The visual evidence can bolster the confidence of decision-makers, making them more inclined to implement data-driven strategies and actions.

Challenges and Considerations in Data Analysis

Data analysis, while a powerful tool in the digital age, is not without its challenges and ethical considerations. These hurdles can significantly impact the integrity of analysis results and the decisions based on them. Under-standing these challenges is crucial for anyone venturing into the field of data analytics.

Data Quality Issues

One of the foremost challenges in data analysis is ensur-ing data quality. Inaccurate, incomplete, or outdated data can lead to misleading analysis results. Common data quality issues include duplicate records, missing values, and inconsistent data formats. Such problems necessitate rigorous data cleaning and preparation processes, yet even the most thorough cleaning cannot always guar-antee the removal of all errors. The adage "garbage in, garbage out" highlights the importance of starting with high-quality data to achieve reliable analysis outcomes.

Data Privacy Concerns

As data analytics often involves handling sensitive or personal information, data privacy emerges as a signifi-cant concern. Adhering to data protection laws, such as the General Data Protection Regulation (GDPR) in the European Union, is mandatory. Analysts must ensure that data is collected, stored, and analyzed in a manner that respects individual privacy and confidentiality. This includes anonymizing data, securing consent for data use, and implementing robust security measures to prevent data breaches.

Ethical Use of Data

Beyond legal compliance, the ethical use of data encompasses broader considerations. It involves using data in ways that are fair, responsible, and beneficial to society. This includes avoiding biases in data collection and analysis that could lead to discrimination or harm to certain groups. Analysts must be vigilant about the potential consequences of their work, striving to use data in ways that contribute positively to society.

The Importance of Critical Thinking and Human Insight

In data analysis, critical thinking and human insight are indispensable. While algorithms and statistical models can process data and identify patterns, they lack the ability to understand context or the nuances of human behavior. Interpreting data analysis results requires a human touch—a deep understanding of the subject matter, the context in which the data was collected, and the potential implications of the findings. Analysts must question assumptions, consider alternative explanations, and recognize the limitations of their analysis.

This need for critical thinking extends to evaluating the sources of data, the appropriateness of the methods used, and the reliability of the conclusions drawn. It involves making informed judgments about which data to trust, which methods to apply, and how to interpret the results in a meaningful way.

The 'Fake It Till You Make It' Phenomenon

Navigating the Shadows

Here, I aim to shed light on a less-discussed aspect of the tech startup world, drawing from my experiences working with both burgeoning and established companies. The realm of technology startups, vibrant and innovative, isn't immune to its share of shadowy practices. Among these, the strategy of "Fake it till you make it" stands out as a contentious path some startups adopt to gain traction and secure funding. My intention isn't to explore the ethical quandaries this strategy presents; instead, I want to explore its implications for data analytics within these organizations.

It's not uncommon for such startups to artificially inflate their user engagement and system activity. This inflation can take various forms, from directly injecting fabricated data into their systems to simulate heightened activity, to employing individuals to create fake accounts and generate artificial interactions. The outcome is an ecosystem brimming with data that blurs the line between genuine user engagement and manufactured interactions.

This practice places startups in a precarious position regarding data analytics. The challenge isn't just in the ethical gray area it occupies but in the tangible impact on the company's ability to generate authentic, actionable insights. The analytics derived from this manipulated data scaffold a narrative of growth that is disconnected from the startup's actual user engagement and market position. While this skewed portrayal might initially satisfy investor expectations, it constructs an unstable foundation for the company's future growth strategies.

The repercussions extend deeply within the company, particularly affecting the product growth team. Tasked with driving genuine user engagement and growth, this team faces monumental challenges when their efforts are shadowed by the artificial inflation of the company's performance metrics. The persistence of this strategy, until a substantial genuine user base is established, casts a prolonged shadow over the startup, where data analytics serve more as a mirror to the illusion of growth than a beacon guiding towards sustainable development.

In essence, the "Fake it till you make it" approach introduces a significant distortion in the startup ecosystem, complicating the pursuit of genuine growth and undermining the integrity of data-driven decision-making. This narrative serves as a cautionary tale, emphasizing the critical need for transparency, authenticity, and ethical practices in leveraging data analytics to shape the trajectory of technology startups.

PART V

Data

in

AI & ML

Data Powers AI

Introduction to AI and Machine Learning

In the tapestry of contemporary technology, artificial intelligence (AI) and machine learning (ML) stand out as two of the most transformative threads, weaving through every aspect of our lives and reshaping the fabric of society. At their core, AI and ML represent the pinnacle of our endeavor to create machines that can think, learn, and adapt—an endeavor that not only challenges our understanding of intelligence but also expands the horizon of human potential.

Defining the Concepts

Artificial Intelligence, in its broadest sense, is the science of making machines capable of performing tasks that typically require human intelligence. These tasks include recognizing speech, making decisions, translating languages, and more. *Machine Learning*, a subset of AI, focuses on the ability of machines to learn from data, identify patterns, and make decisions with minimal human intervention. It is the mechanism through which AI evolves, learns from past experiences, and improves over time.

The Significance of AI and ML

The significance of AI and ML in the modern world cannot be overstated. These technologies drive innovation across sectors, from healthcare, where they enable new diagnostic methods and personalized medicine, to environmental science, where they offer insights into climate change and resource management. In business, AI and ML are revolutionizing industries by automating operations, enhancing customer experiences, and opening new avenues for growth. Their impact extends to our daily lives, powering the smart devices and digital assistants that have become our constant companions.

A Journey Through History

The journey of AI and ML from theoretical concepts to central pillars of the digital age is a story of relentless human curiosity and technological advancement. The seeds of AI were sown in the mid-20th century, with pioneers like Alan Turing questioning the possibility of machines thinking. The subsequent decades saw AI oscillate between periods of optimism and setbacks, known as "AI winters," where progress seemed to stall.

However, it was the exponential growth in data collection and analysis capabilities that truly catapulted AI and ML into the spotlight. The digital revolution, marked by the rise of the internet, smartphones, and sensor technology, generated unprecedented volumes of data. This wealth of information, coupled with advances in computational power and algorithmic design, enabled the development of sophisticated AI and ML models capable of learning and improving at a remarkable pace.

The historical context of AI and ML underscores a journey fueled by data. Each advancement in data collection and analysis techniques has propelled these technologies forward, enabling machines to learn more effectively and tackle increasingly complex tasks. As we continue to generate data at an ever-growing rate, the potential for AI and ML to drive future innovations seems boundless.

As we stand on the cusp of new breakthroughs, understanding the roots and evolution of AI and ML is a window into the future of technology and its role in shaping our world.

The Importance of Data in AI and ML

At the heart of AI and ML lies the concept of learning from experience. For machines, this experience is derived from data. It is through data that AI and ML models gain the capacity to recognize patterns, make decisions, and predict outcomes. The process begins with training, where models are fed vast datasets. These datasets are then used to adjust the model's parameters until it can accurately interpret new data. This iterative process of training and refinement is what enables AI to evolve from basic pattern recognition to understanding complex nuances in data.

The criticality of data in this context cannot be overstated. Without substantial and relevant datasets, AI and ML models lack the necessary material to learn effectively. This limitation can lead to inaccuracies in predictions and decision-making, underlining the adage "quality over quantity." Yet, in the world of AI and ML, the significance of both quality and quantity cannot be overstated. The more diverse and comprehensive the dataset, the more refined and reliable the model becomes.

Types of Data in AI and ML

AI and ML models thrive on a diet of varied data types, each serving different purposes and complexities.

Structured Data

This type of data is highly organized and easily searchable in databases. It includes numerical and categorical data that can be tabulated and queried in a straightforward manner. Structured data is pivotal in training models where clear, definable patterns and relationships need to be understood.

Unstructured Data

Contrary to structured data, unstructured data is not organized in a predefined manner. It encompasses text, images, audio, and video. Processing unstructured data requires more sophisticated AI and ML techniques, such as natural language processing (NLP) and computer vision, to extract meaningful patterns and insights.

The Role of Big Data in AI and ML

Big Data stands as a colossal archive of the human-digital interaction, encapsulating both structured and unstructured data. Its role in developing complex AI and ML algorithms is indispensable. Big Data provides the vast, varied, and continuous stream of information necessary for training robust models. The advent of Big Data has heralded the development of more advanced AI and ML applications, capable of analyzing complex patterns at an unprecedented scale. From predictive analytics in healthcare to real-time fraud detection in finance, the applications are as boundless as the data itself.

Data Preprocessing for ML

Data preprocessing is a crucial step in the Machine Learning (ML) pipeline, ensuring that the dataset is optimized for training models. This process involves several key techniques aimed at improving the quality and efficiency of the data, making it more suitable for analysis. Let's examine the crucial steps involved in preprocessing data to prepare it for a machine learning model.

Data Cleaning

Imagine you're filling out a survey, but you skip a couple of questions. When researchers collect everyone's responses, those skipped questions are like holes in the data. In machine learning, these holes, or missing values, can mess up the results and make the model less accurate.

To fix this, we have a few tricks...

Filling in the Blanks

Sometimes, the easiest solution is to fill in these holes with average values. If we're dealing with numbers, we might use the average (mean), the middle value (median), or the most common value (mode) from all the responses to that question.

Guessing Games

Another approach is to play detective and use clues (other data we have) to make educated guesses about what the missing answers might be. This is where prediction models come into play, helping us estimate those missing values based on patterns in the data.

When There's Too Much Missing

If a survey question was skipped by a lot of people, or if a person skipped a lot of questions, sometimes it's better just to leave out that question or person's responses altogether. This means removing entire rows or columns that have too many missing values.

In simple terms, cleaning data is like patching up holes in a piece of fabric to make sure it's ready and in the best shape for sewing (or in our case, for feeding into a machine learning model).

Dealing with Outliers

Imagine you're in a class and the teacher is handing back test scores. Most scores are between 60 and 90, but there's one score of 150. That 150 is an outlier—it's way different from the rest. In machine learning, outliers are data points that stand out dramatically from the typical pattern. They're like the rogue waves in an otherwise calm ocean, and they can throw a machine learning model off course, leading to misleading results.

So, how do we spot these outliers?

Using a Measuring Tape for Data

One way is by using something called the Interquartile Range (IQR), which is like measuring the heart of the data to see where most values lie. Imagine dividing your data into four equal parts; the IQR is the range between the middle two parts (where most of your data hangs out). Data points that are too far from this cozy middle can be considered outliers.

The Z-score Method

Another method is looking at Z-scores, which tell us how far a point is from the average in terms of standard deviations (a fancy term for measuring how spread out your data is). If a data point is too many standard deviations away (usually more than 2 or 3), it's like saying it's too far from where most of the data is clustering.

Once we find these outliers, we can either remove them or adjust their values to be more in line with the rest of the data. This helps make sure our machine learning model isn't swayed by these anomalies and can learn from the 'normal' pattern of the data.

Data Transformation

Shaping Up Your Data

Think of data transformation as a way to get everyone speaking the same language before they start a group project. It's all about making sure the data is in the right form and scale so that machine learning models can understand and work with it effectively.

Making Data Play Fair

Normalization and Standardization

- **Normalization**
  Imagine you're comparing the heights of your friends to the amount of pocket money they have. These two things are measured in completely different units and scales (inches or centimeters for height, dollars or euros for money). Normalization is like converting everything so it fits within a 0 to 1 scale. It's as if you're resizing everyone's attributes to see them on the same playing field, making it easier to compare apples to oranges, so to speak.

- **Standardization**
  Now, think about lining up all your friends by height, but instead of just resizing, you adjust everyone so the average height is considered the new "middle" (mean of 0), and how much taller or shorter each person is compared to the average

is measured in a unit called "standard deviation." In data terms, this means we're not just squishing the data into a box between 0 and 1; we're shifting and scaling it so that our machine learning models, like support vector machines (SVMs) or neural networks, don't get confused by the wide range of values and can find patterns more smoothly.

Turning Categories into Numbers

Encoding Categorical Data

Let's say you have a survey with a question about favorite colors, and the options are 'Red,' 'Blue,' and 'Green.' These options are categories, not numbers, but machines like numbers, not words. So, we translate these categories into a language machines understand—numbers.

- **One-hot Encoding for Nominal Data**
  Imagine if 'Red,' 'Blue,' and 'Green' are like off/on switches in a row of lights. 'Red' might turn on the first light (1) and leave the others off (0), 'Blue' might turn on the second, and so on. There's no order to these colors; they're just different options, so each one gets its own switch.

- **Label Encoding for Ordinal Data**
  But what if the options show an order, like 'Small,' 'Medium,' and 'Large'? In this case, we can assign numbers directly to these categories in order (like 1, 2, 3). This tells the machine that 'Medium' is between 'Small' and 'Large' without turning everything into separate switches.

By reshaping data through normalization, standardization, and encoding, we essentially tailor the dataset so machine learning models can process, learn from it, and make predictions more efficiently and accurately. It's a crucial step in making sure the data is ready for the big leagues of analysis.

Feature Engineering

Choosing and Creating the Best Ingredients for Your Model

Think of building a machine learning model like cooking a fancy dinner. You start with a variety of ingredients (features) on your kitchen counter (dataset). Not all ingredients will make your dish (model) taste great, and some combinations might work better than others. Feature engineering is about picking the best ingredients and sometimes mixing them in new ways to create a masterpiece.

Picking the Best Ingredients

Feature Selection

Feature Selection is like deciding which ingredients will make your dish stand out. You might have onions, garlic, spices, and more, but for your specific recipe, you need to choose the ones that will truly enhance the flavor. In machine learning:

- *Correlation Matrices* help you see which ingredients (features) complement each other well and which ones just don't mix (for instance, ice cream and ketchup). If two ingredients bring out the same flavors (highly correlated features), you might only need one of them.

- *Backward Elimination* is like tasting your dish as you cook, gradually removing one ingredient at a time to see if the flavor improves. In data terms, you start with all your features and systematically remove the least important one, checking each time if your model performs better without it.

- *Feature Importance Scores from Tree-Based Models* give you direct feedback on which ingredients (features) are making your dish (model) delicious. Like a chef knowing that a pinch of salt is crucial, these scores tell you which features have the most impact on your model's predictions.

Mixing Ingredients in New Ways

Feature Extraction

Feature Extraction is when you mix ingredients to create something even better. Imagine blending herbs and spices to make a perfect seasoning mix.

- *Principal Component Analysis (PCA)* is a technique where you blend features together to make new ones that capture the essence of your data but with fewer components. It's like taking all the individual flavors in your kitchen and combining them into a few signature seasonings that still encompass all the original tastes but in a simpler form.

In simpler terms, feature engineering is about making smart choices with the data you have—selecting the most impactful features and combining them in ways that make your machine learning model work better, just like choosing and mixing the right ingredients can turn a good meal into a great one.

Let's dive into real examples to illustrate the concepts of "Feature Selection" and "Feature Extraction" in a practical context.

Feature Selection

Predicting House Prices

Scenario

Imagine you're building a machine learning model to predict house prices based on a dataset that includes various features: square footage, number of bedrooms, age of the house, distance to the city center, proximity to schools, and color of the house.

Feature Selection Process

- Initially, you include all these features in your model. However, you notice that the color of the house has very little impact on the price. This realization comes from analyzing feature importance scores, which indicate that house color is the least important feature.

- You also observe that both the number of bedrooms and square footage are highly correlated to each other because larger homes tend to have more bedrooms. To simplify your model, you decide to keep only the square footage, as it encompasses the size aspect of the house more comprehensively.

- After removing the color feature and choosing between correlated features, your model becomes more efficient and focused on the variables that truly affect house prices.

Feature Extraction

Customer Segmentation

Scenario

A retail company wants to segment its customers for targeted marketing campaigns. They have customer data including age, annual income, spending score (a measure of purchasing behavior), and the number of transactions per year.

Feature Extraction Process

- The dataset is detailed but complex, making it challenging to directly segment customers. To simplify the analysis, the company uses Principal Component Analysis (PCA) to reduce the dimensionality of the data.

- PCA might combine features like annual income and spending score into a single component that represents a "spending power" dimension, while age and number of transactions per year could be blended into another component representing "shopping frequency and preferences."

- These new components (features) are easier to analyze and provide a distilled view of customer behavior. The company can now segment its customers more effectively, focusing on these simplified, yet informative, features to tailor their marketing strategies.

In the house price prediction example, feature selection helped simplify the model by removing irrelevant information (house color) and choosing between correlated features (square footage over number of bedrooms), enhancing model performance and interpretability.

In the customer segmentation scenario, feature extraction through PCA distilled complex, multi-dimensional data into fewer, more meaningful components, facilitating more efficient and insightful customer segmentation for targeted marketing.

Data Splitting

The Train-Test Split Method

Imagine you're preparing for a big exam. To effectively study, you decide to practice with a set of questions (your entire dataset). But, you also want to make sure you're ready for anything the exam might throw at you, including questions you haven't seen before. To do this, you split your questions into two sets: one set for practicing (training set) and one set to test yourself later to see how well you might perform on the actual exam (testing set).

In the world of machine learning, we do something quite similar with our data, called a train-test split. Here's how it works...

Dividing the Dataset

First, we take our complete collection of data and divide it into two parts. The larger part becomes our training set, which, like your practice questions, we use to teach our machine learning model about the patterns and relationships in the data. The smaller part becomes our testing set, reserved for later, to test how well our model has learned.

Training the Model

We use the training set to 'teach' or 'train' our model, showing it examples of data it can learn from. This is like going through your practice questions, learning the material, and understanding how to solve different types of problems.

Evaluating the Model

After our model has been trained with the training set, we then 'test' it using our testing set. This step is crucial because the testing set contains data that the model hasn't seen before. It's a way to check if the model can apply what it's learned to new, unseen data. This is similar to taking a mock test with questions you've set aside, assessing how well you're prepared for the real exam.

The train-test split is vital for understanding how well our model might perform in the real world. It helps us ensure that our model can generalize its learning to new data, not just the examples it's seen during training. This method gives us confidence in our model's ability to make accurate predictions or decisions when it encounters new data outside of our training set.

Handling Imbalanced Data

Balancing the Scales

Imagine you're playing a game of soccer with friends, but one team has 11 players and the other team only has 5. It's clear which team has a better chance of winning, right? This is similar to what happens in machine learning with imbalanced data. If we're trying to predict two outcomes - let's say, predicting whether emails are spam or not spam - and we have a lot of examples of 'not spam' but very few of 'spam,' our model might get really good at identifying 'not spam' and not so good at spotting 'spam.' It's like it's playing the game with an unfair advantage towards one outcome.

To make the game fair, or in our case, to help the model learn about both classes equally, we use a few strategies...

Oversampling the Minority Class

This is like adding more players to the smaller team to even out the numbers. In data terms, we make more copies of the 'spam' examples so that the amount of 'spam' and 'not spam' data is more balanced.

Undersampling the Majority Class

Going the other direction, this would be like asking some players from the bigger team to sit out, reducing their numbers to match the smaller team. For our data, this means we use fewer of the 'not spam' examples, again aiming for a balance between 'spam' and 'not spam.'

Using Synthetic Data Generation (SMOTE)

Sometimes, just copying data (oversampling) or removing it (undersampling) isn't ideal because it could lead to overfitting or losing valuable information. So, we use a method like SMOTE, which stands for Synthetic Minority Over-sampling Technique. It's like creating new, unique players based on the skills and attributes of the existing players on the smaller team. In data science, SMOTE analyzes the 'spam' emails and generates new, synthetic 'spam' examples that are slightly different but still realistic. This helps balance the data without repeating the same examples over and over or losing important 'not spam' data.

By balancing our data in these ways, we give our model a fair chance to learn about both outcomes - 'spam' and 'not spam' - improving its ability to make accurate predictions in a more balanced and fair manner.

Data Augmentation

Expanding Your Data Wardrobe

Imagine you have a favorite outfit that you love to wear, but you're told you need to show up with a different look every day for a month. You can't buy anything new, so what do you do? You get creative—wear the jacket inside out, flip the scarf to show a hidden pattern, or maybe even wear the belt as a necklace. Essentially, you're taking what you have and tweaking it slightly to create a whole range of outfits from just a few pieces.

Data augmentation does something similar for machine learning, especially in projects that work with images or sequences (like videos or time-series data). When training a model, like one that recognizes objects in photos, you need lots of examples (images) to teach the model effectively. But collecting thousands or millions of images can be tough. So, we use data augmentation to artificially expand our collection of images without actually going out and finding new ones.

Here's how it works...

Rotation

We rotate the images a little to the left or right. If our model can recognize a cat tilted at 15 degrees, it's probably going to recognize one at 20 degrees too.

Flipping

We flip images horizontally or vertically. This way, the model learns that a flipped image of a dog is still a dog.

Adding Noise

We might add a bit of 'noise' or fuzziness to the images. It helps the model learn to ignore irrelevant details and focus on what's important for recognizing the object.

By applying these transformations, we're not just repeating the same images over and over; we're creating new variations that help the model learn better. It's like showing it different 'outfits' of the same object. This process improves the model's robustness, meaning it gets better at recognizing objects in images it has never seen before, even if they're a bit tilted, flipped, or noisy.

In essence, data augmentation is a clever way to make the most out of your existing 'wardrobe' of data, ensuring your model is well-prepared for the real world's variability and diversity.

Final Words

And there we conclude this chapter. I hope to have illuminated the critical role of data and its application in Machine Learning and Artificial Intelligence systems. This journey into the foundational aspects of data is merely the beginning, opening the door to a world of further exploration. As a segment of the Tech Simplified book series, I invite you to dive deeper into the world of AI with titles such as 'AI Simplified,' 'Generative AI Simplified,' and 'NLP Simplified' for those keen on expanding their understanding of Artificial Intelligence and Machine Learning. For easy access to these resources, please scan the QR code below. Let's continue this journey of discovery together.

TechSimplifiedSeries.com

Finally, I truly value your feedback. I encourage you to leave a review for this book and share the insights and discoveries you've made through these pages. Your perspectives not only enrich our learning community but also guide the journey of future readers.

Thank you!

Reader's Notes...

Reader's Notes...

Reader's Notes...

www.ingramcontent.com/pod-product-compliance
Lightning Source LLC
LaVergne TN
LVHW051658050326
832903LV00032B/3875